John Henry Leech

On Lepidoptera Heterocera from China, Japan and Korea

John Henry Leech

On Lepidoptera Heterocera from China, Japan and Korea

ISBN/EAN: 9783742800084

Manufactured in Europe, USA, Canada, Australia, Japa

Cover: Foto ©berggeist007 / pixelio.de

Manufactured and distributed by brebook publishing software
(www.brebook.com)

John Henry Leech

On Lepidoptera Heterocera from China, Japan and Korea

XLIII.—*On Lepidoptera Heterocera from China, Japan, and Corea.* By JOHN HENRY LEECH, B.A., F.L.S., F.Z.S., &c.

[Continued from p. 349.]

Boarmia venustaria. (Pl. VII. fig. 2.)

Boarmia venustaria, Leech, Entom., Suppl. p. 44 (May 1891).

Five specimens, including both sexes, from Oiwake in Pryer's collection.

Hab. Japan.

Boarmia leucophæa.

Boarmia leucophæa, Butl. Ann. & Mag. Nat. Hist. (5) i. p. 395 (1878); Ill. Typ. Lep. Het. iii. p. 33, pl. xlviii. fig. 12 (1879).
Boarmia elegans, Oberth. Etud. d'Entom. x. p. 31, pl. i. fig. 4 (1884).

A very fine series from Yokohama and Oiwake in Pryer's collection, exhibiting considerable variation. My specimens of the male agree exactly with Butler's type of *B. leucophæa* and also with Oberthür's figure of *elegans*; therefore I do not hesitate to consider the latter synonymous with the former.

The female is somewhat smaller than the male and the ground-colour is grey.

Var. *nigrofasciaria*, nov.

Central shade of all the wings black and conspicuous; there is an oblique streak from this to outer margin, and all the transverse lines are very distinct.

Distribution. Japan; Askold.

Boarmia angulifera.

Boarmia angulifera, Butl. Ann. & Mag. Nat. Hist. (5) i. p. 396 (1878); Ill. Typ. Lep. Het. iii. p. 33, pl. xlix. fig. 1 (1879).
Alcis angulifera, ab. *albifera*, Warren, Novit. Zool. i. p. 434 (1894).

There were several specimens from Oiwake and Nikko in Pryer's collection. I obtained the species at Shikotan in August and at Nikko in September; my native collector captured a few examples at Gensan and in the island of Kiushiu in July; Butler's type was from Yokohama; and I received one female specimen from Omei-shan, taken in July.

In some specimens the space between the central lines on primaries is hardly paler than the rest of the wing, in other examples it is almost white.

Distribution. Japan; Corea; Kiushiu; Kurile Islands; Western China.

Boarmia obliquaria.

Hibernia obliquaria, Motsch. Etud. 1860, p. 37.

There was a nice series from Gifu in Pryer's collection and there are specimens from Tokio and Yokohama in the National Museum at South Kensington.

Hab. Japan.

Boarmia mœsta.

Boarmia mœsta, Butl. Trans. Ent. Soc. 1881, p. 407.
Stenotrachelys cinerea, Butl. *op. cit.* p. 409.

There were four specimens in Pryer's collection; three of these are from Oiwake and were placed with *B. conferenda.* The fourth specimen is Pryer's no. 328, which he states was from Fujisan, taken at an elevation of 12,365 feet.

Butler's type of *mœsta* was from Yokohama and his *cinerea* from Tokio.

Hab. Japan.

Boarmia crassestrigata.

Boarmia crassestrigata, Christoph. Bull. Mosc. iv. (2) p. 72 (1881).
Synopsia crassestrigata, Meyrick, Trans. Ent. Soc. Lond. 1892, p. 100.

A few specimens from Yesso in Pryer's collection. I captured some examples at Gensan in June and at Tsuruga in July.

Distribution. Amur; Japan, Yesso; Corea.

Boarmia Büttneri.

Boarmia Büttneri, Hedem. Horæ Soc. Ent. Ross. xvi. p. 54, pl. x. fig. 6 (1881).

I took a nice series at Gensan in June, including one example of the female, which differs from the male in having rather broader wings.

All my specimens are deeper in colour than Hedemann's figure.

Distribution. Amur; Corea.

Boarmia appositaria.

Boarmia appositaria, Leech, Entom., Suppl. p. 46 (May 1891).

A male specimen of this species, which is closely allied to *B. Büttneri*, Hedem., was taken by my native collector at Gensan in July. I have also received two male specimens from Chang-yang and one from Moupin.

Distribution. Corea; Central and Western China.

Boarmia grisea.

Boarmia grisea, Butl. Ann. & Mag. Nat. Hist. (5) i. p. 396 (1878); Ill. Typ. Lep. Het. iii. p. 33, pl. xlix. fig. 2 (1879).

A nice series from Yokohama and Oiwake in Pryer's collection.

I obtained the species at Gensan in July, and at Ningpo I met with it in the month of April.

I have received a male specimen from Kiukiang, taken in May, and a female from Omei-shan, taken in July.

In some specimens the central fascia is very distinct, but in others it is obscured by the ground-colour.

Distribution. Japan; Corea; North-eastern, Central, and Western China.

Boarmia jejunaria, sp. n.

Brownish grey, with slight violet tinge; basal area suffused with ochreous. Primaries have a blackish discal spot surmounted by some blackish scales, and there are three transverse black lines—the first is slightly curved towards costa, the second is obtusely angled below costa and represented by dots on the neuration below the middle; the space between the lines is rather paler than the rest of the wing; submarginal line blackish and wavy, but not distinct. Secondaries have an indistinct black discal spot, a wavy blackish central line, and an indented submarginal line, also blackish; the basal area is freckled with blackish. Fringes of the ground-colour, with black dots at their base between the nervules. Under surface whitish brown; all the wings have a black discal spot, followed by an indistinct transverse line, which does not extend to inner margin on primaries.

Expanse 38 millim.

One female specimen from Ni-tou, July.

Hab. Western China.

Boarmia basifuscaria. (Pl. VII. fig. 14.)

Boarmia basifuscaria, Leech, Entom., Suppl. p. 46 (May 1891).

There was one male specimen in Pryer's collection, and I took an example of the same sex at Oiwake in October.

Hab. Japan.

Boarmia corearia.

Boarmia corearia, Leech, Entom., Suppl. p. 44 (May 1891).

I took four male specimens and one female at Gensan in

July, and I have received one male example from Chang-yang, also taken in July.

Allied to *B. grisea*, Butl., but differs from that species in the non-angulation of the second line.

Distribution. Corea ; Central China.

Boarmia sinuosaria.

Boarmia sinuosaria, Leech, Entom., Suppl. p. 47 (May 1891).

One male specimen taken by myself at Ningpo in April.

Hab. North-east China.

Boarmia definita.

Boarmia definita, Butl. Trans. Ent. Soc. 1881, p. 407.

One male specimen from Oiwake in Pryer's collection. Butler's type was taken by Fenton at Tokio.

Hab. Japan.

Boarmia fuscomarginaria.

Boarmia fuscomarginaria, Leech, Entom., Suppl. p. 45 (May 1891).

One female specimen taken by Mr. Smith at Hakone in August.

Allied to *B. corearia*, but the lines on upper surface are not angulated in the same way and the markings on under surface are different.

Hab. Japan.

Boarmia fumosaria. (Pl. VII. fig. 5.)

Boarmia fumosaria, Leech, Entom., Suppl. p. 44 (May 1891).

Ten specimens, including both sexes, from Oiwake and Yokohama in Pryer's collection.

Hab. Japan.

Boarmia ornataria. (Pl. VII. fig. 15.)

Boarmia ornataria, Leech, Entom., Suppl. p. 45 (May 1891).
Boarmia ornataria, var. *inornataria*, Leech, *l. c.*

One example of the type form taken by a native collector in the island of Kiushiu; there was one specimen of the variety in Pryer's collection.

Hab. Japan and Kiushiu.

Boarmia flavolinearia.

Boarmia flavolinearia, Leech, Entom., Suppl. p. 47 (May 1891).

There were two male specimens in Pryer's collection.

Hab. Japan.

Boarmia montanaria, sp. n.

Primaries brown; first line blackish, slightly curved and preceded by a fuscous transverse shade; second line blackish, angulated below costa, thence oblique to inner margin; between these lines the costal area is filled in with pale brown; submarginal line whitish, preceded by a fuscous shade, which is broadest below the angle of the second line and almost fills up the space between this line and the submarginal; discal spot black and curved, its extremities touching the blackish central shade and forming an annulated mark. Secondaries grey-brown, striated and powdered with darker brown; the central line is indicated by a series of black dots on the nervules; there is an oblique dusky submarginal streak and the anal portion of the indistinct pale submarginal line is also bordered inwardly with dusky. Fringes brown, marked with darker at extremities of the nervules and preceded by a black lunulated line. Under surface greyish, darker on apical area of primaries, and tinged with ochreous on the costal area between the transverse lines; this last and also the discal spot are rather indistinct. Antennæ bipectinated.

Expanse 38 millim.

Seven male specimens from Omei-shan, Ni-tou, and Chetou: July.

Hab. Western China.

Boarmia roboraria.

Geometra roboraria, Schiff. Wien. Verz. p. 101; Hübn. Geom. fig. 169.
Boarmia lunifera, Butl. Ann. & Mag. Nat. Hist. (5) i. p. 395 (1878); Ill. Typ. Lep. Het. iii. p. 32, pl. xlviii. fig. 10 (1879).
Boarmia arguta, Butl. Ann. & Mag. Nat. Hist. (5) iv. p. 372 (1879).
Diastictis roboraria, Meyrick, Trans. Ent. Soc. Lond. 1892, p. 103.

There were specimens from Ohoyama and Yesso in Pryer's collection.

I obtained a specimen at Nagasaki in May, one at Hakodate in August, and one at Nikko in September. Mr. Smith took this species at Hakone in August.

B. lunifera, Butl., is identical with the dark form of *B. roboraria*, which Staudinger has named var. *infuscata*. *B. arguta*, Butl., is also a form of *B. roboraria* in which the transverse markings are well defined and conspicuous.

Distribution. Europe; Amur; Japan; Yesso; Kiushiu.

Boarmia displicens.

Boarmia displicens, Butl. Ann. & Mag. Nat. Hist. (5) i. p. 395 (1878); Ill. Typ. Lep. Het. iii. p. 32, pl. xlviii. fig. 11 (1879).

There were some specimens from Ohoyama and Nikko in

Pryer's collection. Mr. Smith took one at Hakone in August, and I received a male specimen from Ichang, also taken in August.

Distribution. Japan; Central China.

Boarmia consortaria.

Geometra consortaria, Fabr. Mant. Ins. p. 187; Hübn. Geom. fig. 168.
Boarmia conferenda, Butl. Ann. & Mag. Nat. Hist. (5) i. p. 395 (1878); Ill. Typ. Lep. Het. iii. p. 32, pl. xlviii. fig. 8 (1879).
Diastictis consortaria, Meyrick, Trans. Ent. Soc. Lond. 1892, p. 103.

A common insect in Japan and Corea. My collectors met with it in all the localities they visited in Central and Western China.

The Japanese specimens (referable to *conferenda*, Butl.) are generally darker in colour, but they all have the characteristic markings of *consortaria*.

The Chinese specimens are tinged with cinnamon-brown, and some examples are much larger than the type.

Distribution. Europe; Amur; Askold; Corea; Japan; Western and Central China.

Boarmia corticaria, sp. n.

Male.—Pale brown, powdered with dark brown. Primaries clouded with blackish before the whitish submarginal line: first transverse line wavy, second oblique throughout the middle of its course, but dentate below costa and above inner margin, both black; between these lines there is a black spot and a blackish central shade, terminating in a dark brown cloud on inner margin, and beyond the second line there is a dark shade connecting the blackish clouding before submarginal line with a smaller cloud on inner margin; from this shade there are projections along the median nervules. Secondaries have a black central spot, two transverse lines, and a whitish submarginal line; the latter is shaded inwardly with blackish and the second black line is bordered outwardly with dark brown. Fringes brown, preceded by a black lunulated line. Under surface pale brown, irrorated with darker brown; there is a black spot and a broad submarginal fuscous-brown band on each wing; the latter is preceded by a brownish line, which is bifurcated towards abdominal margin and followed by some whitish patches on the outer marginal area. Antennæ bipectinated. Thorax agrees with wings in colour; collar dark brown; tip of abdomen with a tuft of long silky pale brown hairs.

Female.—Rather paler than the male on both surfaces.

Expanse, ♂ 76, ♀ 80 millim.

Three specimens. One example of each sex from Chang-yang and a male from Ichang: June and July.

Hab. Central China.

Boarmia stolidaria, sp. n.

Whitish brown, finely striated with reddish brown. Primaries have a dark irregular basal patch; an obscure dark brown subbasal transverse line, commencing in a darker quadrate patch; beyond there is a curved and recurved dark brown transverse line, clouded on its middle; the submarginal band is twice interrupted. Secondaries have a slightly elongate blackish discal spot, and a serrated dark transverse line, marked with black on the nervules, beyond; submarginal band as on primaries. Under surface similar to above, but the transverse lines of primaries are only faintly indicated. Fringes brownish, tipped with blackish and preceded by a black line.

Expanse, ♂ 64, ♀ 70 millim.

One male specimen from Chang-yang and a female from Ni-tou: July.

Hab. Central and Western China.

Boarmia majuscularia, sp. n.

Ochreous brown, irrorated with fuscous. Primaries have three purplish-brown transverse lines, commencing in clouds of the same colour on the costa; the first is slightly curved, the second is elbowed below costa, and the third is crenulate and undulated and is followed by a large diffuse purplish-brown cloud about the middle; submarginal line whitish, interrupted about the middle—the upper portion is inwardly bordered with purplish brown towards costa, and intersects a large patch of the same colour; the lower portion is also inwardly edged with purplish brown. Secondaries have a brown discal spot and two transverse lines of the same colour; the outer one edged externally with whitish towards abdominal margin. Fringes grey, marked with paler and preceded by an interrupted brown line. Under surface similar to above, but the ground-colour is paler and the transverse markings are only faintly indicated on primaries.

Expanse 74 millim.

One female specimen in Pryer's collection.

Hab. Japan.

Boarmia Pryeraria, sp. n.

Whity brown, tinged with ochreous and marked with

black. Primaries have a subbasal band indicated on costa and inner margin; a large discal spot and two larger spots on outer margin; central line macular and sinuous; submarginal line whitish and sinuous; the space between these lines is clouded with black. Secondaries have a discal dot; the whitish, wavy, submarginal line is broadly bordered inwardly with blackish, and there are some spots on the nervules representing a central line. Fringes of primaries black and of secondaries whity brown preceded by a lunulated line. Under surface whity brown; the outer marginal area of primaries is blackish, enclosing a patch of the ground-colour at apex and another about the middle; the discal spot is distinct; secondaries have a blackish discal spot and submarginal band, the latter encroaching on the outer margin towards apex and again towards anal angle.

Expanse 33 millim.

One female specimen in Pryer's collection.

Hab. Japan.

Boarmia sinicaria, sp. n.

Primaries whitish grey, powdered and suffused with brownish on the costa and on inner marginal area; there is a blackish spot at the base of the wing; basal third suffused with brownish and limited by a double, blackish, nearly straight line; the outer third purplish brown, with an angular projection from the middle to the annular blackish discal spot, and limited by a black angulated line; submarginal line black, marked with whitish on the costa and bordered with violet-grey below; there is a blackish mark on the costa above the discal spot and another beyond it. Secondaries pale whity brown; the outer marginal area is dark grey, and there is a short oblique dash of the same colour above anal angle; discal spot dark grey. Fringes of primaries purplish brown, spotted with yellowish, the central spots are confluent; of secondaries whity brown, slightly darker at the extremity of the nervules; the fringe on all the wings preceded by a series of black lunules. Under surface pale whity brown, costa of primaries yellowish; all the wings broadly bordered with fuscous, and this colour projects on the primaries to the annulated discal spot; the basal two thirds of secondaries are powdered with fuscous, discal spot blackish.

Expanse 36 millim.

One female specimen from Omei-shan, July.

Hab. Western China.

Allied to *B. semiclarata*, Walk.

Boarmia subochrearia, sp. n.

Primaries brownish grey, suffused with fuliginous on outer marginal area; there are three black transverse bands—the first is curved, the second is elbowed below costa and again above the inner margin, the third (submarginal) is sinuous; on the middle of the outer marginal area there is a more or less quadrate pale spot, and the inner margin is tinged with reddish brown; discal dot black. Secondaries ochreous brown, suffused with dark grey towards the base and along the abdominal area; discal dot black. Fringes ochreous brown, marked with dark grey at extremities of the nervules, and preceded by a series of black lunules. Under surface: primaries ochreous grey, with a blackish discal dot and some fuscous clouds beyond it and below apex; of the transverse bands of upper surface only the first is clearly reproduced; secondaries ochreous brown, sprinkled with dark grey; discal dot black.

Expanse 36 millim.

One female specimen from Omei-shan.

Hab. Western China.

Nearly allied to *B. semiclarata*, Walk.

Boarmia bilinearia, sp. n.

Primaries olive-brown, closely striated and mottled with dark brown, traversed by two blackish lines—the first is elbowed below costa, and then curved and recurved to inner margin; the second is strongly serrated; there is a short blackish oblique apical streak. Secondaries greyish brown, mottled with dark brown; discal spot black; transverse line blackish, attenuated from abdominal margin to just beyond middle, where it becomes obscured. Fringes agree with the wings. Under surface pale greyish brown, irrorated with darker; discal spots blackish, and indications of a dusky band beyond the middle of each wing.

Expanse 32 millim.

One male specimen from Moupin, July.

Hab. Western China.

Boarmia punctimarginaria, sp. n.

Fuliginous. Primaries have a pale brown patch at the base limited by a black line; central shade blackish, with the black discal dot on it; beyond there is a black dentate line followed by a blackish shade; on the outer margin there is a series of whitish dots, each dot preceded by a blackish one.

Secondaries similar, but the basal patch is of less extent. Under surface grey-brown; all the wings have a conspicuous black discal spot, and the markings of upper surface are indicated.

Expanse 28 millim.

One male specimen from Kiukiang, June.

Hab. Central China.

Boarmia (?) *nigrofasciaria*, sp. n.

Primaries whity brown; central fascia blackish, tapering towards inner margin, and enclosing a black curved discal spot, the edges dentate; there is a blackish cloud on the costa just beyond the fascia, one on outer margin above the middle, and one towards outer angle; submarginal line pale but indistinct. Secondaries whity brown, with a blackish discal spot and dentate central line. Fringes of the ground-colour preceded by a black line. Under surface similar to above, but markings are fainter. Antennæ bipectinated, the branches rather long.

Expanse 33 millim.

One male specimen from each of the following localities:—Chow-pin-sa, Chia-ting-fu, Pu-tsu-fong: June and July.

Hab. Western China.

Boarmia divisaria, sp. n.

Basal two thirds of primaries brownish, traversed by a blackish line, which is curved towards costa and limited by a dentate black line; outer marginal area brownish, inwardly limited by a waved serrated black line; the area between this line and the basal two thirds is whitish; submarginal line wavy, whitish, preceded by black marks; discal spot black, elongate. Secondaries pale grey, irrorated with fuscous, especially on basal and outer marginal areas; discal dot and central line dusky, the latter almost straight; submarginal line whitish, wavy. Fringes pale brownish grey, preceded by a lunulated brownish line. Under surface ochreous grey, irrorated with fuscous; all the wings have a blackish discal mark and a narrow dusky transverse band beyond, the latter marked with blackish on the neuration. Antennæ bipectinated.

Expanse 36 millim.

One male specimen from Pu-tsu-fong, taken in June or July.

Hab. Western China.

Boarmia decoloraria, sp. n. (Pl. VII. fig. 4.)

Male.—Whity brown, finely powdered with darker brown. Primaries have a brownish basal band (sometimes only represented by a spot on the costa and a dot on each nervure below it); a dark brown serrated central band commencing in an angular mark on the costa, represented by dots on the nervules, and terminating in an oblique dash on the inner margin; submarginal band dark brown, irregular in width and externally edged with whitish, as also is the central band; there is also a brownish spot on costa between basal and central bands, from which a dark shade is sometimes projected to the wide portion of central band. Secondaries have three brownish transverse bands, the first nearly straight, the second narrow and attenuated, and the third of irregular width, the last two edged externally with whitish. All the wings have a blackish discal spot. Fringes brown, preceded by dark brown lunules between the nervules. Under surface of primaries fuscous brown, except the outer and inner margins, which are whity brown; of the secondaries whitish, sprinkled with brown scales; a blackish discal spot on each wing, but only the primaries have the transverse markings, and these are not always clear. Antennæ broadly bipectinated.

Female.—Generally paler than the male, and the under surface of all the wings is usually whity brown, powdered with brown scales (this is also the case in some males); the antennæ are simple.

Expanse, ♂ 40–45, ♀ 36–42 millim.

A long series taken in June and July at Chang-yang; Moupin; Ta-chien-lu; Omei-shan; Wa-shan; Pu-tsu-fong; Chia-ting-fu; Ni-ton.

Hab. Central and Western China.

Near *B. nooraria*, Brem., from Amurland.

Boarmia abietaria.

Geometra abietaria, Hübn. Geom. fig. 160.
Deileptenia abietaria, Hübn. Verz. Schmett. p. 316; Meyrick, Trans. Ent. Soc. Lond. 1892, p. 105.
Boarmia abietaria, Treit. Schmett. vi. 1, p. 204; Dup. Lép. vii. pl. clx. figs. 2, 3; Guen. Phal. i. p. 243.

There were a few specimens from Oiwake in Pryer's collection, and I have received one example of each sex from Mr. Manley, who took them at Yokohama.

The Japanese specimens are rather larger and more strongly marked than European examples.

Distribution. Europe; Japan.

Boarmia approximaria, sp. n.

Allied to *B. abietaria*, Hübn. Greyish, heavily powdered and clouded with brown; the first line of primaries is straighter, the second more deeply elbowed below the costa and more oblique thence to inner margin, where it terminates closer to the first line. The under surface is fuscous and not ochreous as in *B. abietaria*, and the only markings are a discal spot on primaries and indications of a central transverse line on each of the wings.

Expanse, ♂ 48, ♀ 52 millim.

One example of each sex from Ni-tou and a male from Pu-tsu-fong, July.

Hab. Western China.

Boarmia dolosaria, sp. n.

Fuliginous grey. Primaries traversed by two black lines—the first, commencing in a spot on the costa, is sharply angled below costa and again above inner margin; the second is serrated, curved below costa, thence oblique to inner margin. Secondaries have a dusky band with the blackish discal spot on it and a rather sinuous black line beyond; submarginal band dusky, edged outwardly with greyish. Fringes agree with the wings in colour and are preceded by a lunulate black line. Under surface silky fuscous grey; all the wings have a black discal spot, a transverse line, and a broad dusky submarginal band. Antennæ bipectinated.

Expanse 42 millim.

One male specimen from Chang-yang.

Hab. Central China.

Allied to *B. admissaria*, Guen.

Boarmia incongruaria, sp. n.

Male.—Primaries brown; first transverse line darker and only distinct on costal area; second line also darker, slightly elbowed below costa, thence oblique to inner margin; between the lines is a black discal dot and a curved brownish central shade; submarginal line whitish, dentate, preceded by a blackish band. Secondaries rather paler than the primaries; submarginal line pale, bordered inwardly with blackish, but not so strongly as on primaries. Fringes concolorous with the wings. Underside whitish brown slightly tinged with fuscous: primaries have a blackish submarginal band; secondaries a blackish central dot and a whitish transverse line, both indistinct. Antennæ bipectinated.

Female.—Similar to the male, but the secondaries have a blackish central line outwardly edged with whitish, and between this and the base of the wing there is a brownish transverse shade.

Expanse, ♂ 38, ♀ 40 millim.

One example of each sex. The male from Omei-shan and the female from Ni-tou, July.

Hab. Western China.

Boarmia punctilinearia, sp. n.

Pale greyish brown, suffused with darker and striated with blackish. Primaries have a black discal spot and two black transverse lines—the first wavy, preceded by a blackish cloud, and the second serrated, with black dots upon it, and followed on the costa by a clear space of the ground-colour; submarginal band blackish, interrupted. Secondaries have a black discal dot and a serrated black transverse line with black dots upon it; submarginal line as on primaries, but less distinct. Fringes agree with the wings and are preceded by an interrupted black line. Under surface grey, powdered with fuliginous and bordered on outer margins with the same colour; all the wings have a black discal spot and there are indications of one transverse line. Antennæ bipectinated.

Expanse 32 millim.

One male specimen from Huang-mu-chang, July.

Hab. Western China.

Boarmia olivacearia, sp. n.

Dark greyish brown tinged with olivaceous. Primaries have two transverse blackish lines—the first appears to be curved, but is only to be traced from costa to median nervure; second line waved and angled below costa; submarginal line whitish, preceded on the apical area of the wing by a reddish-brown cloud; there is a small black discal spot and a dusky shade beyond extending from the costa to the middle of the wing, the continuation of this shade is represented by a quadrate dusky spot on the middle of the inner margin. Secondaries have a black discal spot and blackish wavy central line; submarginal line as on primaries, the area between the lines is faintly clouded with reddish brown. Fringes grey, marked with brown at the extremities of the nervules, and preceded by a dark dotted line. Under surface brownish grey; all the wings have a blackish waved line, indicated by short dashes on the nervules, a dusky transverse shade between this line and the base of the wing, and a broadish pale band before outer margin. Antennæ bipectinated.

Expanse 40 millim.

One male specimen from Wa-shan, June.

Hab. Western China.

Boarmia projectaria, sp. n.

Primaries blackish grey irrorated with whitish; basal line blackish, indented below costa, and edged internally with whitish; beyond the middle there is a rather broad whitish band forming a double angle about the middle, and outlining in its course a conspicuous quadragular projection of the ground-colour; submarginal line pale but indistinct, with a small blackish cloud on it towards costa; discal spot black. Secondaries whitish, irrorated with blackish grey; discal spot black; central line black, crenulate, with a double-toothed projection about the middle, increasing in width towards abdominal margin. Fringes yellowish, marked with dark grey, and preceded by a series of black spots on the primaries, and by an interrupted black line on secondaries. Under surface similar to above, but the first line is absent.

Expanse, ♂ 28–30, ♀ 33 millim.

Nine specimens, including both sexes, from the following localities, Pu-tsu-fong, Ni-tou, Che-tou, and the Province of Kwei-chow: July.

Hab. Western China.

In some specimens the white band of primaries is suffused with blackish grey, and the secondaries are thickly powdered with the same colour. Antennæ bipectinated in the male.

Boarmia mirandaria, sp. n.

Primaries grey-brown, basal patch pale reddish limited by a curved dark line; a pale reddish patch occupies a large portion of the outer two-thirds of the wing, but does not extend to the costa, and is separated from outer margin by an undulated whitish line, this patch is traversed by a dark wavy line edged with whitish and angled below costa. Secondaries pale reddish, with a patch of grey-brown on the base and on lower portion of the outer marginal area; submarginal line pale; discal spot black. Fringes whitish, preceded by a dark line. Under surface pale brown slightly suffused with fuscous; all the wings have a serrated blackish transverse line beyond the middle, and a dusky shade between it and the base of the wing; secondaries have a blackish discal dot. Antennæ bipectinated.

Expanse 36 millim.

One male specimen from Ichang, June.

Hab. Central China.

Boarmia insolitaria, sp. n.

Primaries have the costal area drab, freckled and clouded with darker brown, and the inner portion of the wing brown, tinged with ferruginous, and clouded with drab on outer margin; there are indications of two dusky transverse lines and a pale submarginal line. Secondaries are drab at the extreme base on outer marginal third, the intervening space being fuscous brown; there is a brownish cloud tinged with ferruginous above anal angle. All the wings have a blackish discal spot, that on primaries only is distinct. Under surface pale brown, clouded and suffused with fuscous; there is a straight dusky central line on which is the discal spot, and beyond there is a slightly curved but indistinct dusky line; the latter is followed by a ferruginous-brown band; the secondaries are similarly marked.

Expanse 42 millim.

One female specimen from Chang-yang, July.

Hab. Central China.

Boarmia moupinaria, sp. n.

Pale grey powdered with darker. Basal half of primaries darker; first line curved, second line almost straight, interrupted in the middle, and bordered with ochreous grey, both are blackish and between them there is a dusky fascia; submarginal line pale grey inwardly edged with blackish. Basal third of secondaries darker; central line blackish, almost straight, and edged with ochreous grey; submarginal line pale grey inwardly edged with darker. Fringes of the ground-colour, preceded by a black line on secondaries and a series of dots on primaries. Under surface whitish grey; costa and outer marginal area of primaries tinged with fuscous; all the wings have a black discal spot.

Expanse 44 millim.

One female specimen from Moupin, June.

Hab. Western China.

Boarmia flavimacularia, sp. n.

Purplish brown. Primaries traversed by two brownish and one silvery-grey line, the latter bordered on each side with reddish brown, and preceded on costal area by a brown and grey suffusion; the base is powdered with silvery grey, and there is a large pale buff apical patch transversely clouded with brownish, and having two black dots on its outer edge. Secondaries have a dusky subbasal line and discal spot, and beyond these are two wavy lines, which become indistinct towards costa: the first of these is inwardly bordered with brownish and has a pale yellow dot on its edge above the middle. Under surface fuscous: primaries have the apical patch as above, and the secondaries are sparingly freckled with pale buff.

Expanse 34 millim.

Five specimens from Pu-tsu-fong, Chia-ting-fu, Chang-yang: July and August.

Hab. Central and Western China.

Genus Jankowskia.

(Oberth. Etud. d'Entom. ix. p. 25 (1884).)

Jankowskia athleta.

Jankowskia athleta, Oberth. Etud. d'Entom. ix. p. 25, pl. ii. fig. 7 (1884).

Recorded by Oberthür from Sidemi, Manchuria.

Jankowskia thoraciaria.

Jankowskia thoraciaria, Oberth. Etud. d'Entom. ix. p. 26, pl. ii. fig. 8 (1884).

Recorded by Oberthür from Sidemi, Manchuria.

Jankowskia fuscaria.

Boarmia fuscaria, Leech, Entom., Suppl. p. 45 (May 1891).

One male and two female specimens from Oiwake in Pryer's collection.

I received the species from Chang-yang, Ichang, Moupin, Omei-shan: June and July.

Allied to *J. athleta*, Oberth., but can at once be distinguished by the yellow marking on the under surface of the costa of primaries.

Distribution. Japan; Central and Western China.

Genus Synopsia.

(Hübn. Verz. Schmett. p. 316.)

Synopsia paupera.

Boarmia paupera, Butl. Trans. Ent. Soc. 1881, p. 406.

Several specimens from Oiwake, Fujisan, and Yokohama in Pryer's collection.

I took the species at Sendai in September, and at Yokohama in October.

Hab. Japan.

Synopsia austeraria, sp. n.

Pale brown, powdered and clouded with darker. Primaries have a black discal dot and two transverse lines—the first is wavy and angulated below costa, the second is slightly serrated, oblique, and preceded by a transverse dusky shade; marginal area clouded with ashy. Secondaries have a blackish discal dot and a blackish serrated transverse line, preceded by a dusky transverse shade; outer marginal area clouded with ashy. Fringes preceded by a dark line. Under surface greyish brown, powdered with darker, and with blackish discal spots and faint indications of transverse lines.

Expanse 49 millim.

One female specimen from Pu-tsu-fong, June.

Hab. Western China.

Genus Hemerophila.

(Steph. Ill. Brit. Haust. iii. p. 189 (1829).)

Hemerophila Dejeani.

Hemerophila Dejeani, Oberth. Étud. d'Entom. x. p. 30, pl. i. fig. 12 (1884); Alph. Rom. sur Lép. vi. p. 60 (1892).

This appears to be a common species in June and July at Ta-chien-lu, Omei-shan, and Pu-tsu-fong. I also received specimens from Ni-tou, Che-tou, and Wa-shan; and Alphéraky records the species from Ou-pin.

It is exceedingly variable in coloration, which ranges from reddish brown to dark olive-brown, and the central area is often very pale.

Hab. Western China.

Hemerophila senilis.

Hemerophila senilis, Butl. Ill. Typ. Lep. Het. ii. p. 48, pl. xxxv. fig. 12 (1878).

Several specimens from Oiwake, Nikko, and Gifu in Pryer's collection.

My native collector captured the species at Hakodate in June or July.

Hampson (Fauna Brit. Ind., Moths, iii. p. 275) considers *H. senilis* to be synonymous with *H. subplagiata*, Walk.

Hab. Japan and Yesso.

Hemerophila conjunctaria, sp. n. (Pl. VII. fig. 9.)

Female.—Allied to *H. senilis*, Butl., but the basal fascia, which, together with the outer marginal area of primaries, is purple-brown in colour, ornamented with lilacine at apex and on inner margin, is broader and more deeply indented below the costa; the central transverse line of secondaries is sinuous, not curved below costa, and there are some lilacine dashes on outer margin; discal spot on all the wings distinct. Under surface whitish, freckled with brownish and a little suffused with dusky on the outer margin of primaries; there is a black discal spot and a dotted line, also black, on each wing. The posterior portion of the thorax is edged with white.

Expanse 46 millim.

Two female specimens from Pu-tsu-fong, July.

Hab. Western China.

Hemerophila atrilineata.

Hemerophila atrilineata, Butl. Trans. Ent. Soc. 1881, p. 405.
Phthonandria atrilineata, Warren, Novit. Zool. i. p. 434 (1894).

There were some examples from Oiwake and Nikko in Pryer's collection, and I captured specimens at Tsuruga in July. My native collector took the species at Hakodate and Gensan, also in July, and I have received one male specimen from Ta-chien-lu, taken in June.

Distribution. Japan; Yesso; Corea; Western China.

Hemerophila rimosa.

Boarmia rimosa, Butl. Ann. & Mag. Nat. Hist. (5) iv. p. 372 (1879).

A few nice specimens from Yokohama in Pryer's collection; my native collector took the species in the island of Kiushiu.

Hab. Japan and Kiushiu.

Hemerophila obscuraria, sp. n.

Reddish brown, much powdered with darker. Primaries have two oblique black lines—the first curves a little as it approaches the inner margin, and the second is wavy throughout and curved towards costa, both are preceded by a blackish shade; there is a dark cloud below the apex and another below it extending to inner margin near the angle, this latter is interrupted. Secondaries have a black line, broad on abdominal margin and tapering towards costa, this is preceded by a dusky transverse shade and followed by two paler bands. At the base of the fringes, which are concolorous with the wings, there is a blackish wavy line. Under surface rather silky brown: primaries have two transverse black lines, chiefly indicated by dots on the neuration, which converge towards the inner margin; secondaries have two equidistant, curved, and wavy black lines.

Expanse 50 millim.

Two specimens from Pu-tsu-fong, June.

Hab. Western China.

Hemerophila latimarginaria, sp. n.

Pale cinnamon-brown, with a black discal spot on all the wings. Primaries have an irregular dark brown line beyond the middle limiting the outer marginal area, which is darker brown and is traversed by a dusky submarginal band; the basal area is slightly irrorated with darker brown. Secondaries have a curved and slightly wavy dark brown central line, the area beyond is darker brown traversed by a dusky submarginal band. Under surface whitish brown, suffused with darker on primaries, and these wings have a blackish discal spot.

Expanse 40 millim.

One female specimen from Ichang, April.

Hab. Central China.

Hemerophila (?) *tachraparia*.

Hemerophila tachraparia, Oberth. Etud. d'Entom. xviii. p. 25, pl. v. fig. 63 (1893).

I have not seen this species. Oberthür describes it from specimens received by him from Ta-chien-lu, Western China.

Genus Medasina.

(Moore, Lep. Ceyl. iii. p. 408 (1886); Hampson, Fauna Brit. Ind., Moths, iii. p. 283 (1895).)

Medasina scotosiaria.

Deinotrichia scotosiaria, Warren, Proc. Zool. Soc. Lond. 1893, p. 420, pl. xxx. fig. 9.
Medasina scotosiaria, Hampson, Fauna Brit. Ind., Moths, iii. p. 284 (1895).

Two male specimens from Pu-tsu-fong, June.

Distribution. Sikhim (*Hampson*); Western China.

Medasina diffusaria, sp. n.

Brown, irrorated with fuscous on secondaries and margins

of primaries. There are indications of two transverse lines on primaries, the outer one represented by blackish dots on the neuration; submarginal band blackish, diffuse and interrupted. Secondaries have a blackish discal spot, two transverse lines, and a diffuse blackish submarginal band, the latter does not extend to costa. Fringes of the ground-colour, preceded by an interrupted blackish line. Under surface rather paler than above; all the wings have a blackish discal spot and an obscure dusky band beyond.

Expanse 74–84 millim.

Four male specimens from Chang-yang, taken in July.

Hab. Central China.

Allied to *M. creataria*, Moore.

Medasina creataria.

Hemerophila creataria, Guen. Phal. i. p. 217 (1857).
Medasina creataria, Hampson, Fauna Brit. Ind., Moths, iii. p. 286 (1895).

Two female specimens from the Province of Kwei-chow, June.

Distribution. Sikhim; Assam (*Hampson*); Western China.

Medasina albidaria.

Boarmia albidaria, Walk. Cat. Lep. Het. xxxv. p. 1582 (1866).
Medasina albidaria, Hampson, Fauna Brit. Ind., Moths, iii. p. 289 (1895).

I received one male specimen from Ichang, April; one example of each sex from Omei-shan; and a female from Moupin, July.

Distribution. Simla; Dharmsála; Sikhim; Khásis (*Hampson*); Central and Western China.

Genus Arichanna.

(Moore, Proc. Zool. Soc. Lond. 1867, p. 658; Hampson, Fauna Brit. Ind., Moths, iii. p. 290 (1895).)

Arichanna tetrica.

Cidaria tetrica, Butl. Ann. & Mag. Nat. Hist. (5) i. p. 451 (1878); Ill. Typ. Lep. Het. iii. p. 59, pl. lv. fig. 10 (1879).
Cidaria tetrica, Butl., ♀; Alph. Rom. sur Lép. vi. p. 78, pl. iii. fig. 11 (1892).

There was an example of each sex from Ohoyama in Pryer's collection, and my native collector took a female specimen at Hakodate in June.

Hab. Japan and Yesso.

Arichanna interruptaria, sp. n.

Primaries whitish, powdered with blackish grey; there are three transverse bands interrupted by the brown neuration and intersected by white or whitish lines; discal spot black, separated from a spot on costa by the brown subcostal nervure; from the lower end of discal spot there is a blackish diffuse line, which appears to be part of the second or central band. Secondaries whitish, powdered and freckled with greyish; discal spot blackish. Fringes pale brown, chequered with darker, preceded by a brown line on secondaries and by a row of black dots on primaries. Under surface pale brown, powdered and freckled with darker: the primaries have a smoky suffusion, a black discal spot, and indications of the transverse bands of upperside; the secondaries have a black discal spot and dark brown wavy central band. Antennæ of the male ciliated.

Expanse 42–46 millim.

Two male specimens and four females from Omei-shan: July and August.

The markings are very similar to those of *A. similaria*, but they are blackish rather than olive-brown, and the neuration is brown instead of pale olive. The structure of the antennæ is quite different.

Hab. Western China.

Arichanna ramosa.

Scotosia ramosa, Walk. Cat. Lep. Het. xxxv. p. 1688 (1866).
Arichanna ramosa, Hampson, Fauna Brit. Ind., Moths, iii. p. 291 (1895).

One female specimen from Pu-tsu-fong, July.

Distribution. Sikhim (*Hampson*); Western China.

Arichanna similaria, sp. n.

Male.—Primaries white, powdered with brownish; the base is dark olive-brown, and there are four transverse bands of the same colour interrupted by the broad pale olive neuration; the second and third of these bands unite on the inner marginal area, where they represent a quadrate patch; the fourth is intersected by an interrupted white line, as also is the third above the quadrate patch referred to. Secondaries whitish, freckled with greyish; central band and incomplete submarginal band greyish; discal spot blackish. Fringes pale olive, chequered with darker. Under surface whitish, powdered and freckled with greyish brown: primaries suffused with shining fuscous except at apex; secondaries have a discal spot and central band as above. Antennæ brown, bipectinated.

Female.—Markings of primaries as in the male; bands of secondaries not well defined beyond abdominal area.

Expanse, ♂ 48, ♀ 50 millim.

Allied to *A. ramosa*, Walk., and *A. tetrica*, Butl.

One example of each sex from Omei-shan, July.

Hab. Western China.

Arichanna clavaria, sp. n.

Male.—Greyish brown, powdered and mottled with darker. Basal area of primaries marked with black and limited by a double-indented black line; discal spot black, preceded by a whitish quadrate patch, and followed by a dusky cloud-like band; on the outer marginal area there is a series of black bars intersected by an interrupted white line; submarginal line white, macular. Secondaries have a dusky discal spot, central band, and interrupted submarginal band. Fringes pale brown marked with darker. Under surface of primaries greyish brown, suffused with blackish, discal spot black; beyond the middle of the wing there are indications of a pale transverse line: secondaries as above. Antennæ bipectinated.

Female.—Similar to the male, but the ground-colour is browner, the whitish patch before discal spot on primaries is absent, and the bands on under surface of secondaries are less distinct.

Expanse, ♂ 42, ♀ 44 millim.

One male specimen from Omei-shan, taken in August, and a female from Pu-tsu-fong, July.

Hab. Western China.

Arichanna Pryeraria.

Arichanna Pryeraria, Leech, Entom., Suppl. p. 51 (May 1891).

I received a male specimen from Mr. Manley of Yokohama, and there was a female example from Oiwake in Pryer's collection.

Allied to *A. furcifera*, Moore.

Hab. Japan.

Arichanna diffusaria, sp. n.

Primaries whitish, with interrupted and irregular fuliginous-brown subbasal, central, and marginal bands, the latter intersected by an oblique streak of the ground-colour from apex; the space between subbasal and central bands is dotted with fuliginous brown, as also is that between central and marginal bands at middle and towards costa and inner margin; costa and nervures marked with ochreous. Secondaries whitish spotted with fuscous grey, the larger of the spots representing central, submarginal, and marginal bands. Under surface similar to above, but the markings, which are pale fuscous on primaries, are fainter, and on secondaries rather stronger.

Expanse 56 millim.

One female specimen from Pu-tsu-fong, June.

Hab. Western China.

Arichanna albomacularia. (Pl. VII. fig. 10.)

Arichanna albomacularia, Leech, Entom., Suppl. p. 51 (May 1891).

Two male and five female specimens, from Gifu and Oiwake, in Pryer's collection.

Allied to *A. tetrica*, Butl., but distinguished by the large white spot on primaries.

Hab. Japan.

Arichanna consocia.

Abraxas consocia, Butl. Ann. & Mag. Nat. Hist. (5) vi. p. 226 (1880).
Icterodes consocia, Butl. Ill. Typ. Lep. Het. vi. p. 84, pl. cxix. fig. 11 (1886).
Arichanna lapsariata, Hampson, Fauna Brit. Ind., Moths, iii. p. 293 (1895).

One female specimen from Ni-tou, July.

Distribution. N.E. Himalayas (*Butler*); Western China.

Arichanna melanaria.

Phal. Geometra melanaria, Linn. Syst. Nat. x. p. 525; Clerck, Icon. pl. iv. fig. 2.
Geometra melanaria, Esp. v. p. 115, pl. xxiii. fig. 1; Hübn. Geom. fig. 26.
Diastictis melanaria, Meyrick, Trans. Ent. Soc. Lond. 1892, p. 104.
Rhyparia fraterna, Butl. Ill. Typ. Lep. Het. ii. p. 53, pl. xxxvii. fig. 9 (1878).
Rhyparia askoldinaria, Oberth. Etud. d'Entom. v. p. 52, pl. ix. fig. 11 (1880).
Icterodes sordida, Butl. Ann. & Mag. Nat. Hist. (5) xi. p. 116.

Fraterna, Butl., is a pale form of *A. melanaria*, and is almost exactly identical with some European examples of the species in my collection. *Sordida*, Butl., is a dark form also agreeing with some European specimens, and *askoldinaria* is a form intermediate between the two. Alphéraky mentions a variety of the species from Peï-chouï (Rom. sur Lép. vi. p. 55).

There were specimens in Pryer's collection from Yokohama, Oiwake, and Nikko. I obtained the species at Gensan in June.

Distribution. Europe; East Siberia; Amur; Askold; Corea; Japan.

Arichanna confusaria, sp. n.

Primaries white; basal area limited by an interrupted blackish band, clouded and marked with the same colour; the central fascia and submarginal line are also blackish, the former encloses spots of the ground-colour and the latter is macular; the spaces between the transverse markings are freckled with blackish. Secondaries whitish, freckled with grey before the blackish, wavy, central band, and ochreous beyond it; submarginal band blackish, broken up into spots, of which that nearest the middle is the largest. Fringes of primaries blackish, chequered with white, and of secondaries yellow, chequered with black, preceded on all the wings by a row of blackish spots. Under surface of primaries have the markings of the upperside indicated, and on the secondaries the markings are reproduced, but the outer half of the wing is only tinged with yellow.

Expanse, ♂ 40–44, ♀ 38 millim.

Two male specimens from Ta-chien-lu and one female from Pu-tsu-fong: June.

Hab. Western China.

In one male the markings on primaries are brownish, but not clearly defined, and the outer margin appears to have a brownish border intersected by a transverse wavy white line.

Arichanna flavovenaria, sp. n.

Male.—Primaries whitish grey suffused with blackish; the basal area is marked with black and limited by a black band intersected by the fulvous venation; discal spot black; submarginal band paler than the ground-colour, followed by a broad black band, which is intersected by the venation and outwardly edged with whitish. Secondaries yellow; basal area fuliginous grey; there are three rows of black spots, those forming the first row being more or less confluent, but not forming an interrupted wavy band as in *A. undularia*. Under surface of primaries fuliginous grey and of secondaries as above, but paler in the ground-colour.

Female.—Similar to the male, but exhibiting more of the ground-colour on primaries above. Under surface of all the wings yellow; primaries flecked and clouded with blackish;

secondaries marked as above, but the basal area is only slightly tinged with fuliginous grey.

Expanse, ♂ 51, ♀ 54 millim.

One male specimen from Omei-shan and a female from Pu-tsu-fong: July.

Hab. Western China.

Arichanna flavomacularia, sp. n.

Primaries black; venation and costa broadly grey, the latter marked with five black spots, and the former, together with seven interrupted yellow lines, breaking up the ground-colour into macular transverse bands. Secondaries have the basal third grey from costa to anal angle and the outer two thirds yellow; the latter has a central black spot, an outer series of six large grey spots (the first and sixth of which are double), and a marginal series of seven or eight small spots of the same colour. Fringes of primaries blackish and of secondaries yellow to the anal third, where they are blackish. Under surface of primaries fuliginous grey, with black spots of upperside reproduced; secondaries as above.

Expanse, ♂ 56–60, ♀ 54 millim.

A fine series from Wa-shan and Ta-chien-lu, June and July. All but one are males.

Hab. Western China.

Arichanna undularia, sp. n.

Male similar to *A. flavomacularia*, but smaller, and the interrupted yellow transverse lines on primaries are less clearly defined. On the secondaries there is a rather broad transverse black band in addition to outer and marginal series of spots; the blackish basal third does not extend beyond the limit of waved band, but encroaches further along the costa and encloses the discal spot. Fringes of all the wings blackish. Under surface: primaries fuliginous grey, with a black discal spot and a few yellowish dots beyond; secondaries marked as above, but the ground-colour is paler.

Expanse 53 millim.

Four male and two female specimens from Ta-chien-lu, Omei-shan, Pu-tsu-fong, Wa-shan: June and July.

Hab. Western China.

Arichanna lateraria, sp. n.

Primaries grey, with a slight fuliginous tinge; the basal area is spotted with black; a conspicuously large black spot with black cloud above on costa, and four pairs of spots of the same colour below, represent a broad central fascia (the lower pairs coalesce, forming bars); beyond these are three transverse series of black spots—the first is composed of double spots towards costa, the spots of second series are surrounded with whitish, the third series is on outer margin. Secondaries have the basal half grey and the outer half yellow; two large black spots beyond the central one form a longitudinal series of three between the limit of basal half of the wing and the outer margin; there is a larger black spot at the outer angle, three others below the outer one of longitudinal series, and a row of smaller spots before the outer margin. Fringes agree in colour with the wings. Under surface similar to the upper surface, but the primaries are more uniform in colour.

Expanse 60 millim.

Three males, Wa-shan, Moupin, and Pu-tsu-fong: July.

Hab. Western China.

Allied to *A. jaguararia*, Guen.

Judging from the three specimens under observation, this species would seem to be rather variable in number of black markings on secondaries, as in one example there is a fourth spot above those of marginal series, and in another specimen, which also has this fourth spot, there are, in addition, two spots above and two below the middle one of longitudinal series of three, thus forming a macular band.

Arichanna jaguarinaria.

Rhyparia jaguarinaria, Oberth. Etud. d'Entom. vi. p. 17, pl. ix. fig. 1 (1881).

Arichanna jaguarinaria, Hampson, Fauna Brit. Ind., Moths, iii. p. 295 (1895).

One male specimen from Wa-shan, June.

Oberthür's type was from the province of Kwei-chow.

Hab. Western China.

This is probably only a form of *R. jaguararia*, Guen., in which the central macular band of secondaries is either obsolete or only faintly indicated; the yellow on these wings does not extend much beyond the submarginal series of spots.

Arichanna jaguararia.

Rhyparia jaguararia, Guen. Phal. ii. p. 198 (1857).

Several specimens from Ohoyama and Oiwake in Pryer's collection. I took the species at Tsuruga in July, and I have received it from Kiukiang and Ningpo. Guenée's type was from N. China, and there are specimens in the National Collection at South Kensington from Yokohama, Hakone, Tokio, and Ashi-no-yo.

The specimens from Oiwake differ from the typical form in having the ground-colour of primaries and basal area of secondaries whiter, and for this form I propose the varietal name *pallidaria*.

Distribution. Japan; Central and Northern China.

Arichanna Gaschkevitchii.

Boarmia Gaschkevitchii, Motsch. Bull. Mosc. xxxix. p. 197 (1866).

Probably this species is identical with *Arichanna* (*Rhyparia*) *jaguararia*, Guen.

Arichanna flavomarginaria.

Rhyparia flavomarginaria, Brem. Lep. Ost-Sib. p. 83, pl. vii. fig. 11 (1864).
Abraxas flavomarginaria, Græser, Berl. ent. Zeit. 1888, p. 390.
Diastictis flavomarginaria, Meyrick, Trans. Ent. Soc. Lond. 1892, p. 104.

Several specimens were captured by my native collector at Gensan in July. I have also received the species from Chang-yang, Moupin, and the province of Kwei-chow: June and July.

Distribution. East Siberia; Amur; Corea; Central and Western China.

The specimens in my series exhibit considerable variation in the size, number, and intensity of the black markings.

Genus Erebomorpha.

(Walk. Cat. Lep. Het. xxi. p. 494 (1860).)

Ereboniorpha consors.

Ereboniorpha consors, Butl. Ill. Typ. Lep. Het. ii. p. 52, pl. xxxvii. fig. 3 (1878).
Mesastrape consors, Warren, Novit. Zool. i. p. 432 (1894).

A few specimens from Fujisan and Nikko in Pryer's collection.

My native collector obtained the species at Hakodate in June or July, and there are specimens in the National Collection from Yokohama and Tokio. I have also received examples from Moupin and Chang-yang, July.

Distribution. Japan; Yesso; Western and Central China.

Some of the Chinese specimens appear to be intermediate between *E. consors*, Butl., and *E. fulgurita*, Walk.

Genus PHYLLABRAXAS, nov.

Palpi porrect, reaching slightly beyond the frons and clothed with long hair; outer margins of wings rounded. Primaries of male with fovea. Neuration as in *Arichanna*. Antennæ of male ciliated and finely serrated. Hind tibiæ dilated, with a tuft of long hair and two pairs of moderately long spurs.

Type *P. curvaria*.

Phyllabraxas curvaria, sp. n. (Pl. VII. fig. 3.)

Whitish, sparingly powdered with brownish. Primaries suffused with olive-grey and marked with fuliginous brown as follows:—a spot at the base; an angulated subbasal band united at the angle with a central fascia, the latter encloses a patch of the ground-colour at each end and its outer edge is curved and recurved; on the outer margin below apex there is a more or less quadrate spot, and on the costa before apex there is a similar one (these spots are united at their opposed angles), the lower encloses two white dots, which represent portions of a much interrupted submarginal line; on the outer marginal area above outer angle there is a third spot, and from the inner edge of this a streak descends to inner margin. Secondaries have a blackish discal spot and a submarginal band, the latter indicated by a brown spot on the costa, one about the middle, and a dash above anal angle. Fringes of primaries black, chequered with whitish towards inner margin; those of secondaries whitish marked with blackish. Under surface: primaries whitish, the apical spots are blackish and the basal area is suffused with the same colour; secondaries are whitish powdered with blackish, and have an incomplete central band in addition to the macular submarginal band as above, but both are blackish.

Expanse 42–46 millim.

Five specimens, including both sexes, from Ta-chien-lu, Moupin, and Omei-shan: July.

Hab. Western China.

In one female from Ta-chien-lu (the only specimen from that locality) the apical markings are browner.

Phyllabraxas similaria, sp. n.

Primaries sordid whitish, freckled with blackish and marked with reddish brown as follows:—a small patch at the base; a straight narrow subbasal fascia, commencing as a spot on costa; an outwardly diffuse central fascia, broadest on costa, and enclosing a patch of the ground-colour at each end; a rather quadrate patch on costa before apex and a similar one on outer margin below apex, united at opposed angles; from the inner edge of the lower patch a straight narrow fascia descends to inner angle, but is only clearly defined below the second median nervule; discal spot black. Secondaries white, freckled with pale grey; discal spot blackish; central and submarginal fasciæ, the former hardly darker than the freckling, the latter represented by a blackish spot on costa, another about the middle, and a short blackish bar above anal angle. Fringes pale grey-brown, preceded on primaries by a row of black dots, becoming lunular towards inner angle, and on secondaries by a thin brownish line. Under surface: primaries fuscous grey, tinged with ochreous on costa, markings of upperside faintly reproduced; secondaries whitish, with the markings as above, but the central fascia is rather darker and the submarginal less distinctly indicated.

Expanse 40–42 millim.

Two male specimens from Pu-tsu-fong and one from Omei-shan: July.

Hab. Western China.

Allied to *P. curvaria*, but it differs in size and colour and also in the form of the fascia.

Phyllabraxas exsoletaria, sp. n.

Primaries whitish grey, speckled with black and suffused with smoky grey at the base of the wing and on outer marginal area; central fascia broad, slightly olivaceous, its inner edge black and undulated and its outer edge black and obtusely angled below costa; before the black discal spot there is a short black line from the costa, this is connected by a blackish suffusion with the external edge of fascia, thus forming a more or less quadrate patch on the outer costal portion of the fascia; the short line referred to has a dusky continuation to the inner margin, but it is not clearly defined; beyond the angle of fascia there are some black dots, and some other dots are placed towards inner margin and parallel with edge of fascia; submarginal line whitish, interrupted. Secondaries rather smoky white, with a black discal spot and indications of a central line. Fringes grey, preceded on the primaries by black dots between the nervules. Under surface smoky grey; basal area of primaries limited by a pale undulated and exteriorly diffuse band; secondaries greyish, speckled with darker; discal spot black; there is another black spot about centre of the wing and blackish suffusion along abdominal margin.

Expanse 45 millim.

Three male specimens from Pu tsu-fong, June and July.

Var. *divisaria.*

The primaries are suffused with brownish; the central fascia is divided transversely into two parts, the outer being dark brown and the inner paler brown. Secondaries whitish, with discal spot and dusky central line which swells out into a spot about the middle and towards each extremity.

Expanse 43–48 millim.

Two male specimens from Omei-shan, July.

Hab. Western China.

Phyllabraxas conspicuaria, sp. n.

Primaries white, with olivaceous markings; basal patch pale olive-brown; central fascia clouded with darker olive-brown, enclosing white discal spot and limited outwardly by an oblique series of black dots; the inner edge is slightly curved below costa; outer marginal area clouded and suffused with dark olive-brown, limited inwardly by a pale olive-brown band and enclosing two diffuse white spots. Secondaries white, freckled with greyish; incomplete central band, discal spot, and shade between the latter and abdominal margin darker grey. Fringes grey and whitish, preceded by an interrupted blackish line. Under surface: primaries have the basal two thirds dusky, limited by a series of darker dots and enclosing a white discal spot; outer area dusky, with a whitish spot at apex and another about the middle: secondaries are whitish, freckled with grey-brown; central band blackish, macular; discal spot, a small cloud on abdominal margin, and three spots on costa also blackish.

Expanse 39 millim.

Three male specimens from Pu-tsu-fong, June.

Hab. Western China.

Genus ABRAXAS.

(Leach; Hampson, Fauna Brit. Ind., Moths, iii. p. 297 (1895).)

Abraxas evanescens.

Callabraxas evanescens, Butl. Trans. Ent. Soc. 1881, p. 420.

A fine series from Oiwake and Yesso in Pryer's collection. My native collector took the species at Hakodate in August.

Hab. Japan and Yesso.

Abraxas placida.

Abraxas placida, Butl. Ann. & Mag. Nat. Hist. (5) i. p. 441 (1878); Ill. Typ. Lep. Het. iii. p. 48, pl. liii. fig. 1 (1879).
Callabraxas propinqua, Butl. Trans. Ent. Soc. 1881, p. 420.

A few specimens from Oiwake, Nikko, and Yesso in Pryer's collection. I captured several examples at Hakodate in August.

Hab. Japan and Yesso.

Placida, Butl., appears to be an aberrant form of *propinqua*, Butl., but as it was the first to receive a name, it must stand as the type of the species, and *propinqua*, which is really the normal form, must rank as a variety. This is certainly unfortunate, but is not by any means a singular case.

Abraxas Whitelyi.

Abraxas Whitelyi, Butl. Ill. Typ. Lep. Het. ii. p. 52, pl. xxxvii. fig. 4 (1878).

A nice series from Oiwake, Yesso, and Nikko in Pryer's collection.

I captured specimens at Gensan in June, and my native collector at Hakodate in June or July. I have also received two specimens from Mr. Manley, of Yokohama.

There is a good deal of variation in the size of the black markings. In some of the specimens, including all the examples from Oiwake, the large spots on costa and inner margin, representing the central band, are not intersected by the ground-colour as in the type. These specimens have also distinct macular submarginal and marginal bands on all the wings.

Distribution. Amur (*Græser*); Corea; Japan; Yesso.

Abraxas languidata.

Abraxas languidata, Walk. Cat. Lep. Het. xxiv. p. 1122 (1862).
Callabraxas languidata, Hampson, Fauna Brit. Ind., Moths, iii. p. 518 (1895).

Four specimens from Ohoyama in Pryer's collection.

I captured a specimen at Shimonoseki in July, and have received two examples from Omei-shan, also taken in July.

Distribution. Japan and Western China.

Abraxas martaria.

Abraxas martaria, Guen. Phal. ii. p. 205 (1857); Hampson, Fauna Brit. Ind., Moths, iii. p. 300 (1895).

I took a specimen at Foochou in April, and have received the species from Kiukiang, Ta-chien-lu, and Moupin: June.

This species may be distinguished from all the forms of *A. sylvata* by its larger size, more intense dark markings, and by the almost uninterrupted dark costal border.

Hampson considers *A. pusilla*, Butl., to be a small form of *A. martaria*, Guen., and the latter as possibly an extreme form of *A. sylvata*.

Distribution. Nepal; Sikhim; Bhután; Assam (*Hampson*); Eastern, Western, and Central China.

Abraxas sylvata.

Phalæna sylvata, Scop. Ent. Carn. p. 220 (1763).
Zerene leopardina, Koll. Hüg. Kasch. iv. p. 490.
Abraxas sylvata, Hampson, Fauna Brit. Ind., Moths, iii. p. 299 (1895).
Abraxas miranda, Butl. Ann. & Mag. Nat. Hist. (5) i. p. 441 (1878); Ill. Typ. Lep. Het. iii. p. 48, pl. lii. fig. 12 (1879).
Abraxas suffusa, Warren, Novit. Zool. i. p. 417 (1894).
Abraxas latifasciata, Warr. *l. c.* p. 419.
Abraxas fulvobasalis, Warr. *l. c.*
Abraxas suspecta, Warr. *l. c.*
Abraxas deminuta, Warr. *l. c.*
Abraxas degener, Warr. *l. c.*

This species appears to vary in Eastern Asia to even a greater extent than in Europe. I have specimens from various localities in Japan and from all the localities visited by my collectors in Central and Western China. Among these are examples agreeing more or less exactly with the forms named above, together with others that are intermediate between such forms and the more typical specimens. The largest individual in the series measures 60 millim. in expanse and the smallest 31 millim.

Distribution. Europe; Amur; Japan; Corea; Central and Western China.

Abraxas concinna.

Abraxas concinna, Warren, Novit. Zool. i. p. 421 (1894).

Described from Thibet. My collectors did not meet with this species.

Abraxas grossulariata.

Phal. Geometra grossulariata, Linn. Syst. Nat. x. p. 525.
Abraxas conspurcata, Butl. Ill. Typ. Lep. Het. iii. p. 48, pl. lii. fig. 11 (1879).
Abraxas flavisinuata, Warren, Novit. Zool. i. p. 420 (1894).

There were four examples of the *conspurcata* form from Oiwake and two of the *flavisinuata* form from Fujisan in Pryer's collection.

In the form *conspurcata* the markings on secondaries are

certainly more decided than in any European specimen of *grossulariata* that I have seen, but the pattern is only a complete development of markings seen more or less clearly indicated in the majority of European *grossulariata.* Neither of the Japanese forms of this species diverge so widely from the type as do certain varieties of the species known to British entomologists.

I have received a nice series from Chang-yang, taken in July. In these specimens the markings on primaries are somewhat similar to those of *A. picaria*, Moore, but the markings on secondaries are much the same as in typical *A. grossulariata*, though not so pronounced—fuscous instead of black on all the wings. The yellow markings are in all cases less distinct, and in several specimens entirely obsolete. I propose the name *sinicaria* for this form.

Distribution. Europe; Siberia; Amur; Japan; Central China.

Abraxas picaria.

Abraxas picaria, Moore, Proc. Zool. Soc. Lond. 1867, p. 652.

Appears to be a common species in Western China, occurring in July and August.

Specimens of the typical form exhibit considerable modification in the amount of black on primaries; in some examples this colour largely predominates. In addition to what may be regarded as ordinary aberration, there are three forms of the species from Western China, each of which appears to be worthy of a distinctive name.

Var. *tortuosaria*, nov.

In this form the only prominent markings on the primaries are the costal portions of subbasal line, the discal spot, the sinuous and deeply angled transverse line beyond the middle, and a series of spots on outer margin; the secondaries are only sparsely dotted with fuscous on basal area, but the other markings are much as in the type.

Ta-chien-lu, Omei-shan, Moupin: July.

Var. *griscaria*, nov.

Primaries whitish, heavily clouded and spotted with grey; the base is yellow, edged with black; discal spot black, with a black cloud-like spot before it on the costa, from the last there is sometimes a blackish shade extending to inner margin; submarginal band blackish, mixed with yellow, elbowed just above the middle. Secondaries whitish, spotted with blackish, especially on abdominal area, which is tinged with yellowish; discal spot blackish; a yellow band dotted or edged with black extends from just above anal angle to a little beyond third median nervule. Fringes grey on primaries, whitish on secondaries, preceded on each wing by a row of black spots. Under surface similar to the upperside, but the transverse markings are not distinct on primaries and there is a black spot on the middle of the costa of secondaries.

Expanse 44–46 millim.

Nine male specimens from Pu-tsu-fong, July.

Var. *nebularia*, nov.

Primaries whitish, mottled and clouded with smoky brown; discal spot blackish, placed in the lower end of a cloud on costal area; basal area marked with yellow; beyond the middle of the wing there is a yellow irregular line, but this is not clearly defined. Secondaries whitish, sprinkled with brownish; there is a small cloud-like spot on the middle of the abdominal margin, and a short dash of yellow, bordered on each side by brownish spots, above anal angle; central spot blackish. Fringes brownish, marked with whitish at ends of the nervules. Under surface similar to above, but the yellow markings of primaries are absent and there are some yellow hairs at the base of secondaries.

Expanse 44–54 millim.

I have a long and rather variable series of this form, the specimens comprised therein being from Ta-chien-lu, Pu-tsu-fong, Ni-tou, Wa-shan, Omei-shan, and Chia-ting-fu: July.

Distribution. Kumaun; Sikhim (*Hampson*); Western China.

Abraxas punctisignaria, sp. n. (Pl. VII. fig. 13.)

Primaries pale yellowish buff, marked with yellow at the base of costa and sparingly spotted with black; the most conspicuous of these spots are three or four on basal portion of costa and a transverse series beyond the middle of the wing, the spots forming the lower portion of the series are rather larger than the others and are placed on a yellow abbreviated band from the inner margin. Secondaries paler than primaries, but more liberally spotted with black, the central series followed on abdominal margin by a yellow patch; all the wings have a black discal spot. Fringes pale yellowish buff. Under surface similar in colour to that of the upperside of the secondaries; black markings as above, but there is no trace of yellow. Body yellow, marked with black on the back and sides.

Expanse 40 millim.

Two male specimens: one from Moupin, July, and one from the summit of Omei-shan, August.

Hab. Western China.

Abraxas flavobasalis, sp. n.

Male.—Creamy white; basal area of primaries yellow, spotted with black and limited by a black macular line elbowed at costa; some blackish spots beyond the discal spot form fairly regular central and submarginal series, the latter outwardly bordered towards the inner margin with yellow; marginal series much interrupted and not well defined. Secondaries have a blackish spot at the base and some other spots of the same colour arranged in three transverse series, the middle series bordered with yellow towards anal angle. Under surface whitish, spots as above, but no yellow markings. Body yellowish, marked on the sides of abdomen with black and also on the dorsal surface of the seventh, eighth, and ninth segments.

Female.—Similar to the male, but the ground-colour is rather whiter.

Expanse, ♂ 36, ♀ 39 millim.

One example of each sex from Chang-yang, July.

Hab. Central China.

Abraxas punctaria, sp. n.

Female.—Yellowish buff. Basal area of primaries yellow, dotted with black and limited by three black dots, one on the median and submedian nervures respectively and one between the costa and subcostal nervure; beyond the discal spot there are four transverse series of black dots, with a short yellow band between the second and third, starting from inner margin. Secondaries have four transverse series of black dots, with a yellow abbreviated band between the second and third, as on primaries; immediately preceding the innermost of these series there are some diffuse dots above the abdominal margin. Under surface whiter than above; transverse series of dots hardly so distinct; no yellow markings. Head and thorax yellow, dotted with black; abdomen pale yellowish, with a black dot on the back and side of each segment except the terminal one and that next the thorax.

Expanse 32 millim.

One female specimen from Moupin, July.

Hab. Western China.

Abraxas reticularia, sp. n.

Primaries white, sparingly dotted with brownish; traversed by two diffuse blackish-brown transverse bands and a diffuse longitudinal band of the same colour; the outer of the transverse bands is bifurcate on the costa and expands on the inner margin, sometimes it is intersected by a thin whitish line; above the black discal spot there is a blackish cloud, from which a spur descends to longitudinal band. Secondaries white, speckled with brownish and exhibiting traces of a brownish central band; discal spot blackish. Fringes white, more or less chequered with brownish. Under surface similar to the upper. Body yellow, marked with black.

Expanse 40–44 millim.

Five male specimens and one female from Ta-chien-lu, Omei-shan, Ni-tou, and Che-tou: July.

Hab. Western China.

Abraxas curvilinearia, sp. n. (Pl. VII. fig. 12.)

White. Primaries sparingly freckled with greyish brown; basal patch dark ochreous, bordered with brown; discal spot brown, surmounted by a brownish cloud on costa; there is a transverse curved and recurved brown band beyond the middle, expanding towards inner margin. Secondaries have a series of five brownish spots, terminating in a brownish bar, outwardly bordered with yellow, on abdominal margin. Under surface: colour as above, markings faintly reproduced.

Expanse, ♂ 44, ♀ 46 millim.

One example of each sex from Chia-ting-fu, July.

Hab. Western China.

Genus Ligdia.

(Guen. Phal. ii. p. 209 (1857).)

Ligdia japonaria, sp. n. (Pl. VII. fig. 1.)

White. Basal area of primaries smoky brown, spotted with black and limited by a curved series of black spots; above the anal angle there is a smoky-brown patch, intersected by the median nervules and connected by a narrow blackish band with a blackish blotch on the costa; discal spot blackish, with a spot above it on the costa and one below it on inner margin, and there are some spots and clouds of the same colour on the outer marginal area. Secondaries have a blackish central spot and broad transverse band, and there are some marks of the same colour on abdominal margin and also on the outer margin. Fringes greyish white, marked with darker on primaries. Under surface similar to above, but the basal area of primaries is clouded with blackish and all the markings are of the same colour.

Expanse, ♂ 28, ♀ 31 millim.

Several specimens in Pryer's collection from Oiwake.

Hab. Japan.

Allied to *L. adustata*, Schiff.

Ligdia ciliaria, sp. n.

White. Primaries have a blackish basal patch and border on outer margin, the latter is broadly interrupted in the middle; discal spot black. Secondaries have a black discal spot and the outer margin is bordered as on primaries; there is a blackish cloud on the middle of the abdominal margin; all the dark patches are suffused with golden brown. Fringes golden brown, marked with a rather darker shade. Underside similar to above, but the borders are not quite so broad.

Expanse 28 millim.

One female specimen from Oiwake in Pryer's collection.

Hab. Japan.

Also allied to *L. adustata*, Schiff.

Genus LOMASPILIS.

(Hübn. Verz. Schmett. p. 316.)

Lomaspilis marginata.

Phal. Geometra marginata, Linn. Syst. Nat. x. p. 527; Clerck, Icon. pl. ii. fig. 5.
Abraxas marginata, Meyrick, Trans. Ent. Soc. Lond. 1892, p. 116.
Lomaspilis opis, Butl. Ann. & Mag. Nat. Hist. (5) i. p. 442 (1878); Ill. Typ. Lep. Het. iii. p. 49, pl. liii. fig. 3 (1879).

There were several specimens from Oiwake, Nikko, and Yesso in Pryer's collection. I have also received the species from Chang-yang, June.

Most of the Japanese examples of *L. marginata* in my series are referable to the form which Butler has named *opis*, but some are very typical.

Græser also records var. *opis* from Amurland (Berl. ent. Zeit. 1888, p. 391).

Distribution. Europe; Amur; Japan; Yesso; Central China.

Genus METABRAXAS.

(Butl. Trans. Ent. Soc. Lond. 1881, p. 419.)

Metabraxas clerica.

Metabraxas clerica, Butl. Trans. Ent. Soc. Lond. 1881, p. 419.
Metabraxas clerica, var. *inconfusa*, Warren, Novit. Zool. i. p. 415 (1894).

There were a few specimens from Oiwake and Yesso in Pryer's collection.

I captured several examples at Hakodate in August, and have received one female from Chang-yang, where it was taken in July. Butler's type was from Tokio.

The black spots are variable in size and degree of confluency; in var. *inconfusa*, Warr., they are well separated.

Distribution. Japan; Yesso; Central China.

Metabraxas luridaria, sp. n.

White. Basal area of primaries leaden grey, marked with ochreous; costa broadly marked with leaden grey; central fascia, submarginal and marginal bands also leaden grey, the first two macular, the second united with the third on apical area and also towards inner margin. Secondaries have the following leaden-grey markings:—a spot on median nervure, with one between it and a spot on abdominal margin; a central fascia, the middle portion of which is broken up into twin spots; a macular submarginal and a marginal band united as on primaries. Fringes grey. Under surface as above, but the base of the primaries is not marked with ochreous. Antennæ of the male ciliated. Head brownish grey, face whitish. Thorax light brown, marked with darker. Abdomen grey, with two black spots on each segment above. Legs grey.

Expanse 54 millim.

One male example from Moupin, July.

Hab. Western China.

In general appearance this species greatly resembles *M. rufonotaria*, but the antennæ of the male are different in structure and the arrangement of the markings, although very similar, is not identical.

Metabraxas rufonotaria, sp. n.

White, with leaden-grey and brownish markings. On the primaries the leaden-grey markings comprise a basal patch and a streak along the costa, a broad central fascia interrupted just above the middle; a submarginal band, also interrupted above the middle, and only separated from the marginal band by a transverse row of spots of the ground-colour; between the basal patch and central fascia there is a short interrupted dash from the costa; the brownish marks are placed on the basal patch and costal portion of short dash beyond and also on the costal and inner marginal portions of the central fascia. Secondaries have a blackish discal spot and a smaller spot on the first fork of median nervure; the leaden-grey central fascia is interrupted about the middle and sometimes before inner margin; marginal and submarginal bands as on primaries, but the latter is not marked with brown. Fringes dark grey in the male, paler in the female. Under surface as above, but there are no brownish markings on primaries. Antennæ of the male bipectinated. Head greyish, face brownish. Thorax grey, marked with brown. Abdomen yellow, with two black spots on each segment above. Legs yellow, tarsi marked with black.

Expanse, ♂ 54, ♀ 56 millim.

Four specimens (three males and one female) from Omei-shan, July.

Hab. Western China.

Metabraxas incompositaria, sp. n.

Male.—White, spotted and marked with dingy grey. On the primaries these markings represent a basal patch and central fascia, with a short band between them from costa to discal spot, and a broad band on outer marginal area, the last is composed of smaller spots. Secondaries have an interrupted central band, a discal spot, and some smaller spots between it and the base of the wing; some scattered spots on outer marginal area. Fringes grey, narrowly interrupted with white on the secondaries and towards the inner angle of primaries. Under surface: markings as above, but blacker. Antennæ of the male bipectinated. Head and thorax yellow, spotted with black, face blackish. Abdomen yellowish. Legs grey, marked with blackish.

Female.—Similar to the male, but the spots on basal and inner marginal areas are smaller and those on outer marginal area fewer in number, especially on the secondaries.

Expanse 62 millim.

Seven male specimens and one female from Chang-yang, June.

Hab. Central China.

Metabraxas molossaria.

Abraxas molossaria, Oberth. Etud. d'Entom. x. p. 32, pl. iii. fig. 10 (1884).

This species is described by Oberthür from Tong-Tchéou (province of Kwei-chow). My collectors did not meet with it in any part of China that they visited.

Oberthür states that this is a variable species and that he has received a melanic form of it from Northern India.

Hab. Western China.

Metabraxas (?) *nigromarginaria*, sp. n.

White, with broad fuliginous borders to all the wings; these borders are preceded by a series of spots of the same colour; on the primaries there are, in addition, a broad fuliginous streak along the costal portion of the basal third of the wing and some yellow markings at the extreme base; discal spot black, with a short interrupted band beyond; the apical portion of the marginal border of primaries is very broad and encloses some spots of the ground-colour; the inner marginal area of these wings is rather thickly spotted with fuliginous. Secondaries have a few dark spots at their base. Fringes of primaries fuliginous, slightly marked with white; of secondaries white, chequered with fuliginous. Under surface as above. Antennæ bipectinated nearly to the tip. Body black, marked with yellow.

Expanse 54 millim.

One male specimen from Wa-shan, June.

Hab. Western China.

Genus DILOPHODES.

(Warren, Novit. Zool. i. p. 416 (1894).)

Dilophodes elegans.

Abraxas elegans, Butl. Ill. Typ. Lep. Het. ii. p. 53, pl. xxxvii. fig. 6 (1878).

Dilophodes elegans, Warren, Novit. Zool. i. p. 416 (1894); Hampson, Fauna Brit. Ind., Moths, iii. p. 305 (1895).

Several specimens from Ohoyama, Nikko, and Gifu in Pryer's collection.

I captured the species in Satsuma in May, and Mr. Smith at Hakone in August. One male specimen was obtained by my collectors in the province of Kwei-chow, a female example at Omei-shan, and several specimens at Chang-yang: July.

The black markings of the Japanese specimens are larger and more confluent than in Chinese examples.

Distribution. Japan; Western and Central China. Khásis (*Hampson*).

Dilophodes conspicuaria, sp. n.

White, marked with black. Primaries spotted on basal area; central fascia interrupted, broadest on costa; outer marginal area banded with black; this band, which is narrowest in the middle, is intersected with white or whitish along the neuration and traversed by a white line; between the central fascia and marginal band is a spot on costa, sometimes united with the former. Secondaries have a spot at the base and one between it and the discal spot; there is also a spot on the middle of abdominal margin; outer marginal area as on primaries. Fringes of primaries black, tending to grey towards the inner angle; those of secondaries are pale grey. Under surface similar to the upperside, but there is a broad black dash along the costa of secondaries extending from basal spot to outer marginal band. Head and thorax ochreous, the latter spotted above with black and the patagia marked with whitish; abdomen white, dorsally marked with black, and the terminal segment with a tuft of long silky grey hairs.

Expanse ♂ 58–64, ♀ 52–60 millim.

There were two male specimens and four females from Gifu in Pryer's collection. I have also received two males from Central China, where they were captured at a place thirty miles north-west of Ichang in July; in one of these the line traversing the outer marginal border is obsolete.

Distribution. Japan; Central China.

Genus Percnia.

(Guen. Phal. ii. p. 216 (1857).)

Percnia foraria.

Percnia foraria, Guen. Phal. ii. p. 217 (1857); Leech, Trans. Ent. Soc. Lond. 1889, p. 146; Hampson, Fauna Brit. Ind., Moths, iii. p. 307 (1895).
Xenoplia foraria, Warren, Novit. Zool. i. p. 415 (1894).

There were a few specimens from Yokohama in Pryer's collection.

I received the species from Chang-yang, Kiukiang, Omei-shan, and the province of Kwei-chow: June and July.

Distribution. Dharmsála; Simla; Sikhim (*Hampson*); Japan; Central and Western China.

Percnia belluaria.

Percnia belluaria, Guen. Phal. ii. p. 217 (1857).
Percnia guttata, Feld. Reis. Nov. v. pl. cxxx. fig. 15, ♂ (1874).
Percnia belluaria, Hampson, Fauna Brit. Ind., Moths, iii. p. 308 (1895).

Three specimens from Chang-yang, one from Wa-shan, four from Pu-tsu-fong, and one from Omei-shan: June and July. I took this species in September in Southern Kashmir and I have received it from Kulu.

Distribution. Sikhim; Khásis (*Hampson*); Kulu; Kashmir; Central and Western China.

Percnia grisearia, sp. n.

Male.—Whitish. Primaries tinged with cinnamon-grey at the base and clouded with blackish on basal half; all the wings are traversed by series of black dots arranged as in *P. fumidaria*, but the bands are blackish grey and considerably interrupted, especially on the lower portion of outer margin of primaries and on the secondaries. Fringes whitish grey, becoming darker towards apex of primaries. Under surface of primaries much clouded with blackish grey, that of secondaries very similar to upperside, but the markings are rather paler. Thorax and abdomen cinnamon-grey, spotted with black.

Female.—Wings rather more ample; outer margin of primaries rounder. The blackish-grey central band of primaries is much broken up and the bands on secondaries are almost obsolete; the under surface of primaries is free from blackish-grey clouding except on basal portion of costa and on apical area of primaries.

Expanse, ♂ 58, ♀ 60 millim.

Ten specimens (seven males, three females) from Kiukiang, Ichang, Chang-yang, Kwei-chow, and Chia-ting-fu.

Hab. Central and Western China.

This species and also *P. fumidaria* are allied to *P. belluaria*, Guen.

Percnia fumidaria, sp. n.

Male.—Whitish. Primaries suffused with pinky grey on the basal half and traversed by five rows of black dots, those of central and two inner series placed on the neuration and those of outer series placed between the nervules; the central series is followed by a broad band of pinky grey and the space between the two outer series is grey, with the exception of some bars of the ground-colour between the opposed dots;

the discal spot is black and there are two black dots at the base of wing. On the secondaries the arrangement of black dots and pink-grey bands corresponds with that on primaries, except that there are only two black dots and the discal spot between the central series and the base of the wing. Fringes pinky grey. Under surface whitish; discal spot on all the wings black, larger than above and followed by a transverse series of short black dashes on the nervules; two outer series of dots as on upperside, but those of primaries obscured by a band of grey.

Female.—Similar to the male, but the wings are rather more ample, and the outer margin of primaries rounder; the pinky-grey suffusion is more restricted to basal area.

Expanse, ♂ 46, ♀ 52 millim.

Several examples of each sex from Chang-yang and Ichang and a pair from Chia-ting-fu: July.

Hab. Central and Western China.

Percnia giraffata.

Abraxas giraffata, Guen. Phal. ii. p. 205 (1857).
Rhyparia grandaria, Feld. Wien. ent. Mon. 1862, p. 39; Reise Nov. pl. cxxix. fig. 28 (1874).
Percnia giraffata, Hampson, Fauna Brit. Ind., Moths, iii. p. 309 (1895).

Two specimens from Fujisan in Pryer's collection. Three specimens from Chang-yang, one from Wa-shan, two from Omei-shan, one from Moupin, and one from the province of Kwei-chow: June.

Distribution. Sikhim and Burma (*Hampson*); Japan; Central and Western China.

Percnia exanthemata.

Culcula exanthemata, Moore, Lep. Atk. p. 266 (1887).
Percnia exanthemata, Hampson, Fauna Brit. Ind., Moths, iii. p. 308 (1895).
Buzura abraxata, Leech, Trans. Ent. Soc. Lond. 1889, p. 143, pl. ix. fig. 14, ♀.

I received specimens of this species from Kinkiang, Moupin, and Omei-shan, all taken in July.

The males range from 64–72 millim. in expanse and the females from 74–90 millim.

Distribution. Sikhim; Khásis (*Hampson*); Central and Western China.

Genus OBEIDIA.

(Walk. Cat. Lep. Het. xxiv. p. 1139 (1862).)

Obeidia vagipardata.

Obeidia vagipardata, Walk. Cat. Lep. Het. xxiv. p. 1139 (1862).

Common at Chang-yang, Kiukiang, and in all the localities visited by my collectors: June and July.

There is considerable variation in the size of the black spots; those on the secondaries are often confluent and form more or less complete bands. In one male specimen from the province of Kwei-chow the secondaries are almost entirely blackish and the outer and inner marginal areas of primaries are broadly marked with the same colour.

Hab. Central and Western China.

Obeidia rongaria.

Rhyparia rongaria, Oberth. Etud. d'Entom. xviii. p. 35, pl. ii. fig. 22 (1893).

Described by Oberthür from a specimen received by him from Tsé-kou.

Hab. Western China.

Obeidia idaria.

Rhyparia idaria, Oberth. Etud. d'Entom. xviii. p. 35, pl. v. fig. 73 (1893).

Oberthür's type was from Tsé-kou.

Hab. Western China.

Obeidia tigrata.

Abraxas tigrata, Guen. Phal. ii. p. 202 (1857).
Obeidia tigrata, Hampson, Fauna Brit. Ind., Moths, iii. p. 309, fig. (1895).
Rhyparia leopardaria, Oberth. Etud. d'Entom. vi. p. 17, pl. ix. fig. 5 (1881).

A large number of specimens were received from Moupin, Omei-shan, Chia-ting fu, Kiukiang, and Chang-yang: June and July. I captured the species at Gensan in July.

This species varies in expanse, in the tone of the yellow coloration, and also in the size of the black spots; in some specimens the spots on secondaries are confluent and form bands.

Leopardaria, Oberth., is certainly a form of this species;

I have a specimen from Moupin which is almost exactly identical with Oberthür's figure.

Distribution. Sikhim; Nágas; Penang (*Hampson*); Corea; Central and Western China.

Obeidia gigantearia, sp. n.

Male.—Yellow, central area of all the wings white; basal, costal, and marginal areas heavily spotted with black. All the wings have a black irregular fascia and a diffuse submarginal band: there is also a more or less complete subbasal band on the secondaries. Fringes yellow, chequered with black. Under surface as above.

Female.—Similar to the male, but the central fascia and submarginal band on all the wings are broken up into spots, as also is the subbasal band on secondaries.

Expanse 86–92 millim.

A large number of specimens from the province of Kwei-chow and from Omei-shan and Moupin, also one example from Chang-yang: June and July.

Hab. Central and Western China.

Obeidia conspurcata, sp. n.

Male.—Yellow, central area of all the wings white; basal, costal, and marginal areas spotted with blackish; all the wings have a broad central fascia and an ill-defined submarginal band, the former composed of large blackish spots and the latter of smaller spots. Fringes yellow, chequered with blackish. Under surface as above.

Female.—Similar to the male, but the wings are rather more ample and the spots forming the central fascia are smaller and more scattered.

Expanse 70–72 millim.

A long series from Chang-yang, also several specimens from Omei-shan, Moupin, and Kwei-chow: July.

Hab. Central and Western China.

This species is very like *O. gigantearia*, but it is smaller and the maculation is not so black or so heavy, and the bands are never so well defined.

Obeidia (?) *Largeteaui.*

Rhyparia Largeteaui, Oberth. Etud. d'Entom. x. p. 32, pl. i. fig. 5 (1884).

Appears to be common at Ichang, Chang-yang, and Omei-shan. I also received one specimen from Chia-ting-fu: June and July.

Hab. Central and Western China.

Obeidia (?) aurantiaca.

Halthia aurantiaca, Alph. Rom. sur Lép. vi. p. 56, pl. iii. fig. 2, ♂ (1892).

Alphéraky records a male specimen from the river Heï-hò, in the province of Kan-Sou: July.

Genus Vithora.

(Walk. Cat. Lep. Het. iv. p. 818 (1855).)

Vithora stratonice.

Phalæna stratonice, Cram. Pap. Exot. iv. p. 234, pl. cccxcviii. fig. K.
Cystidia stratonice, Hübn. Verz. Schm. p. 174 (1800).
Vithora agrionides, Butl. Ann. & Mag. Nat. Hist. (4) xv. p. 137 (1875); Ill. Typ. Lep. Het. pt ii. p. 3, pl. xxii. fig. 3 (1878).
Vithora stratonice, Leech, Proc. Zool. Soc. Lond. 1888, p. 614.
Cistidia stratonice, Meyrick, Trans. Ent. Soc. Lond. 1892, p. 116.

I met with this species at many places in Japan during the months of May, June, and July; at Gensan in June; and I have received two specimens from Kiukiang.

Distribution. Japan; Corea; Central China.

Vithora indrasana.

Vithora indrasana, Moore, Proc. Zool. Soc. Lond. 1865, p. 795, pl. xlii. fig. 5.
Halthia nigripars, Swinh. Trans. Ent. Soc. Lond. 1892, p. 16, pl. i. fig. 1.
Vithora indrasana, Hampson, Fauna Brit. Ind., Moths, iii. p. 311 (1895).

I have a long series of this species from Moupin and Omei-shan, captured in July.

Distribution. Sikhim; Khásis (*Hampson*); Western China.

The specimens agree better with the form *nigripars* from Khásis than with Sikhim specimens. The central white markings on secondaries, however, are smaller, especially the dash in cell, which is represented by a patch in the cell and a small spot just outside.

Vithora conaggaria.

Abraxas conaggaria, Guen. Phal. ii. p. 202 (1857).
Halthia eurypyle, Mén. Bull. de l'Acad. Pét. xvii. p. 217; Schr. Amur-Reise, p. 47, pl. iv. fig. 3 (1859).
Halthia eurymede, Motsch. Etud. d'Ent. 1860, p. 30.
Cistidia conaggaria, Meyrick, Trans. Ent. Soc. Lond. 1892, p. 116.
Abraxas interruptaria, Feld. Wien. ent. Mon. 1862, p. 39; Reise der Nov. pl. cxxix. fig. 29; Leech, Trans. Ent. Soc. Lond. 1889, p. 145.
Abraxas lithosiaria, Walk. Cat. Lep. Het. xxiv. p. 1125 (1862).

There were several specimens in Pryer's collection, and I

captured some fine examples at Nagahama and Gensan in July. I have also received the species from several Chinese localities.

Distribution. Japan; Corea; Amur; Central and Western China.

The very extensive series that I have retained comprises all the intergrades between a specimen which is white in ground-colour, with narrow black bands, and one which is black in colour, with three small white spots on the basal area and two spots, rather larger, on the outer third of primaries. The secondaries of this specimen are black, with two white bands.

Genus Neolythria.

(Alph. Rom. sur Lép. vi. p. 72 (1892).)

Neolythria djrouchiaria.

Abraxas djrouchiaria, Oberth. Etud. d'Entom. xviii. p. 34, pl. iii. fig. 37 (1893).

I received a number of specimens from Ta-chien-lu, Moupin, and Che-tou. In some of these the black transverse band of primaries is not intersected by a yellow line.

Var. *montana*, nov. (Pl. VII. fig. 11.)

Smaller than the type; the transverse band of primaries is broader, the black spots on outer margin of each wing are united and form a marginal band, the secondaries have a distinct macular submarginal band.

A long series was taken on the summit of Mount Omei in August.

In one female example of this form the ground of primaries is yellowish, and the secondaries are tinged with the same colour.

Hab. Western China.

Neolythria abraxaria.

Neolythria abraxaria, Alph. Rom. sur Lép. vi. p. 72, pl. iii. figs. 8 *a, b* (1892).

This species was first discovered in the province of Szechuen. The specimens I have received from Western China do not agree with the type; I therefore describe them as var. *confinaria*, nov.

In this form the white submarginal band on primaries is much narrower and its edges are serrated; the lower discal streak is shorter, and there is sometimes a small triangular spot between this and the upper streak; on the secondaries the marginal black spots are larger.

Three specimens from Che-tou and two from Ta-chien-lu, July.

Hab. Western China.

Neolythria tandjrinaria.

Abraxas tandjrinaria, Oberth. Etud. d'Entom. xviii. p. 34, pl. ii. fig. 23 (1893).

Occurs not uncommonly at Chin-kou-ho, Wa-shan, Huang-mu-chang, and Chang-yang: June and July.

Among the specimens from Chin-kou-ho was an example of the female, and as this sex has not been previously described, I append a short description.

Female.—Yellow band, together with the black internal border, on outer margin of primaries much narrower than in the male; the inner row of spots only represented by two or three on the costal portion. On the secondaries the spots of the inner row on outer margin are linear and almost touching the outer ones.

Hab. Western and Central China.

Neolythria consimilaria, sp. n.

Similar to *A. tandjrinaria*, Oberth., but the black border of the yellow band on outer margin of primaries, which is rather fulvous in tint, is deeply indented on costal area; the black spots forming the inner series are larger, especially the costal one and that below it; the two rows of black spots on outer margin of secondaries are wider apart; all the wings have a distinct black discal spot.

Expanse 30–32 millim.

Several specimens (all males) from Wa-ssu-kow; examples have also been received from Ta-chien-lu and Pu-tsu-fong: June and July.

Hab. Western China.

Neolythria Oberthüri, sp. n. (Pl. VII. fig. 6.)

Also similar to *A. tandjrinaria*, Oberth., but the outer marginal band of primaries is not tapered, but of almost uniform width from costa to inner margin; the inner row of black spots is free and does not unite with the black bordering line at any point. On the secondaries the marginal series of black spots are even wider apart than in *A. consimilaria*, and there is a black dot on each nervule before the inner series;

the black discal spot of primaries is linear and slightly curved, that on secondaries, when present, is punctiform.

Expanse, ♂ 36, ♀ 38 millim.

Seven male specimens and three females from Moupin and Huang-mu-chang: July and August.

Hab. Western China.

Neolythria nubiferaria, sp. n.

White, slightly tinged with smoky, venation blackish. Primaries: costa blackish, most broadly so near the white transverse line, which precedes a deep blackish border on outer margin; discal spot black. Secondaries have a small black discal spot. Fringes whitish, preceded by a series of blackish lunules. Under surface similar to above.

Expanse, ♂ 26, ♀ 28 millim.

One example of each sex from How-kow.

Hab. Thibet.

Genus Xanthabraxas.

(Warren, Novit. Zool. i. p. 422 (1894).)

Xanthabraxas hemionata.

Abraxas hemionata, Guen. Phal. ii. p. 208 (1857).
Xanthabraxas hemionata, Warren, Novit. Zool. i. p. 422 (1894).

I received four specimens from Chang-yang, one from Kiukiang, and five from Moupin: July.

Hab. Central and Western China.

EXPLANATION OF THE PLATES.

Plate VI.

Fig. 1. *Heterocallia truncaria*, sp. n., p. 212.
Fig. 2. *Urapteryx subpunctaria*, Leech, p. 192.
Fig. 3. —— *similaria*, sp. n., p. 192.
Fig. 4. *Anonychia praeditaria*, sp. n., p. 226.
Fig. 5. *Xyloscia biangularia*, sp. n., p. 210.
Fig. 6. *Oberthüria nigromacularia*, sp. n., p. 189.
Fig. 7. —— *flavomarginaria*, sp. n., p. 188.
Fig. 8. *Heterolocha quadraria*, sp. n., p. 231.
Fig. 9. *Rumia trimacularia*, sp. n., p. 298.
Fig. 10. *Krananda lucidaria*, sp. n., p. 305.
Fig. 11. *Psychostrophia picaria*, sp. n., p. 189.
Fig. 12. *Venilia flavaria*, sp. n., p. 233.
Fig. 13. *Myrteta sinensaria*, sp. n., p. 195.
Fig. 14. *Macaria elongaria*, sp. n., p. 308.
Fig. 15. *Pericallia marmorataria*, sp. n., p. 207.

PLATE VII.

Fig. 1. *Ligdia japonica*, Leech, p. 449.
Fig. 2. *Boarmia venustaria*, Leech, p. 414.
Fig. 3. *Phyllabraxas curraria*, sp. n., p. 411.
Fig. 4. *Boarmia decoloraria*, sp. n., p. 424.
Fig. 5. ——*fumosaria*, Leech, p. 417.
Fig. 6. *Neolythria Oberthüri*, sp. n., p. 461.
Fig. 7. *Selenia adustaria*, Leech, p. 205.
Fig. 8. *Biston emarginaria*, sp. n., p. 322.
Fig. 9. *Hemerophila conjunctaria*, sp. n., p. 431.
Fig. 10. *Arichanna albomacularia*, Leech, p. 436.
Fig. 11. *Neolythria djrouchiaria*, Oberth., var. *montana*, nov., p. 460.
Fig. 12. *Abraxas curvilinearia*, sp. n., p. 449.
Fig. 13. —— *punctisignaria*, sp. n., p. 447.
Fig. 14. *Boarmia basifuscaria*, Leech, p. 416.
Fig. 15. —— *ornataria*, Leech, p. 417.

XLIV.—*Descriptions of Two new* Muridæ *from Central and West Africa.* By W. E. DE WINTON.

THE examination of some specimens of West-African Muridæ lately acquired by the British Museum and kindly entrusted to me for determination by Mr. Oldfield Thomas shows the necessity for descriptions being drawn out and names given to two forms. One I propose to name *Mus sebastianus*, the peculiarity of the fur suggesting arrows sticking in its skin. The other I name *Malacomys centralis*; the examples of this species were collected and presented to the British Museum by Dr. Emin Pasha ten years ago, and referred to by Thomas (P. Z. S. 1888, p. 11) as *M. longipes*, but have until now never been compared with specimens of *M. longipes*, M.-Edw. The Museum has since acquired several specimens of this West-African form.

Mus sebastianus, sp. n.

Size rather smaller than *M. rattus*: whole of the upper parts dull coffee-brown, fur soft and rather woolly, interspersed with long shining lance-shaped darker hairs; beneath greyish white, not sharply separated from the colour of the upper parts; feet and hands covered with fine short adpressed brown hairs; nails pale horn-colour, small on the fore feet, those on the hind feet much larger and stronger, curved, but not very sharp; front part of the face and nose thickly haired; whiskers all black-brown, long, reaching well beyond ears; in the alcoholic specimens the ear laid forward just reaches to postcanthus of the eye; tail very long, unicoloured dark slate, smooth and practically naked.

www.ingramcontent.com/pod-product-compliance
Lightning Source LLC
Chambersburg PA
CBHW022024190326
41519CB00010B/1589

* 9 7 8 3 7 4 2 8 0 0 0 8 4 *